NATIONAL ACADEMIES Sciences Engineering Medicine

NATIONAL ACADEMIES PRESS
Washington, DC

Pathways to an Equitable and Just Energy Transition

Principles, Best Practices, and Inclusive Stakeholder Engagement

---

Anne Frances Johnson, *Rapporteur*

Committee on Accelerating Decarbonization in the United States: Technology, Policy, and Societal Dimensions

Board on Energy and Environmental Systems

Division on Engineering and Physical Sciences

Board on Atmospheric Sciences and Climate

Division on Earth and Life Studies

Board on Environmental Change and Society

Division of Behavioral and Social Sciences and Education

Transportation Research Board

Proceedings of a Workshop

NATIONAL ACADEMIES PRESS 500 Fifth Street, NW Washington, DC 20001

This study was supported by the Alfred P. Sloan Foundation, Gates Ventures, Heising-Simons Foundation, Incite Labs, and Quadrivium Foundation with support from the National Academy of Sciences' Andrew W. Mellon Foundation Fund, National Academy of Sciences' Arthur L. Day Fund, and National Academy of Sciences' Thomas Lincoln Casey Fund. Any opinions, findings, conclusions, or recommendations expressed in this publication do not necessarily reflect the views of any organization or agency that provided support for the project.

International Standard Book Number-13: 978-0-309-70176-1
International Standard Book Number-10: 0-309-70176-7
Digital Object Identifier: https://doi.org/10.17226/26935

This publication is available in limited quantities from:

Board on Energy and Environmental Systems
500 Fifth Street, NW
Washington, DC 20001
bees@nas.edu
http://www.sites.nationalacademies.org/DEPS/BEES

This publication is available from the National Academies Press, 500 Fifth Street, NW, Keck 360, Washington, DC 20001; (800) 624-6242 or (202) 334-3313; http://www.nap.edu.

Printed in the United States of America.

Suggested citation: National Academies of Sciences, Engineering, and Medicine. 2023. *Pathways to an Equitable and Just Energy Transition: Principles, Best Practices, and Inclusive Stakeholder Engagement: Proceedings of a Workshop.* Washington, DC: The National Academies Press. https://doi.org/10.17226/26935.

The **National Academy of Sciences** was established in 1863 by an Act of Congress, signed by President Lincoln, as a private, nongovernmental institution to advise the nation on issues related to science and technology. Members are elected by their peers for outstanding contributions to research. Dr. Marcia McNutt is president.

The **National Academy of Engineering** was established in 1964 under the charter of the National Academy of Sciences to bring the practices of engineering to advising the nation. Members are elected by their peers for extraordinary contributions to engineering. Dr. John L. Anderson is president.

The **National Academy of Medicine** (formerly the Institute of Medicine) was established in 1970 under the charter of the National Academy of Sciences to advise the nation on medical and health issues. Members are elected by their peers for distinguished contributions to medicine and health. Dr. Victor J. Dzau is president.

The three Academies work together as the **National Academies of Sciences, Engineering, and Medicine** to provide independent, objective analysis and advice to the nation and conduct other activities to solve complex problems and inform public policy decisions. The National Academies also encourage education and research, recognize outstanding contributions to knowledge, and increase public understanding in matters of science, engineering, and medicine.

Learn more about the National Academies of Sciences, Engineering, and Medicine at **www.nationalacademies.org**.

Consensus Study Reports published by the National Academies of Sciences, Engineering, and Medicine document the evidence-based consensus on the study's statement of task by an authoring committee of experts. Reports typically include findings, conclusions, and recommendations based on information gathered by the committee and the committee's deliberations. Each report has been subjected to a rigorous and independent peer-review process and it represents the position of the National Academies on the statement of task.

Proceedings published by the National Academies of Sciences, Engineering, and Medicine chronicle the presentations and discussions at a workshop, symposium, or other event convened by the National Academies. The statements and opinions contained in proceedings are those of the participants and are not endorsed by other participants, the planning committee, or the National Academies.

Rapid Expert Consultations published by the National Academies of Sciences, Engineering, and Medicine are authored by subject-matter experts on narrowly focused topics that can be supported by a body of evidence. The discussions contained in rapid expert consultations are considered those of the authors and do not contain policy recommendations. Rapid expert consultations are reviewed by the institution before release.

For information about other products and activities of the National Academies, please visit www.nationalacademies.org/about/whatwedo.

COMMITTEE ON ACCELERATING DECARBONIZATION IN THE UNITED STATES: TECHNOLOGY, POLICY, AND SOCIETAL DIMENSIONS

STEPHEN W. PACALA (NAS), Princeton University, *Chair*
DANIELLE DEANE-RYAN, Bezos Earth Fund
ALEXANDRA FAZELI, National Association of State Energy Officials
JULIA H. HAGGERTY, Montana State University
CHRIS T. HENDRICKSON (NAE), Carnegie Mellon University
ADRIENNE L. HOLLIS, Hollis Environmental Consulting
ROXANNE JOHNSON, BlueGreen Alliance
TIMOTHY C. LIEUWEN (NAE), Georgia Institute of Technology
VIVIAN E. LOFTNESS, Carnegie Mellon University
CARLOS E. MARTÍN, Brookings Institution and Harvard University
MICHAEL A. MÉNDEZ, University of California, Irvine
CLARK A. MILLER, Arizona State University
JONATHAN A. PATZ, University of Wisconsin–Madison
KEITH PAUSTIAN, Colorado State University
WILLIAM PIZER, Resources for the Future
ED RIGHTOR, American Council for an Energy-Efficient Economy
PATRICIA ROMERO-LANKAO, National Renewable Energy Laboratory and University of Chicago
DEVASHREE SAHA, World Resources Institute United States
KELLY SIMS GALLAGHER, Tufts University
SUSAN F. TIERNEY, Analysis Group
WILLIAM WALKER, University of California, Berkeley

Staff

K. JOHN HOLMES, Board Director/Scholar, Board on Energy and Environmental Systems (BEES), *Study Director*
ELIZABETH ZEITLER, Associate Director, BEES
STEPHEN GODWIN, Scholar, Transportation Research Board Consensus and Advisory Studies Division
DAVID BUTLER, J. Herbert Hollomon Scholar and Director, Cultural, Ethical, Social, and Environmental Responsibility in Engineering Program, National Academy of Engineering
ELIZABETH FINKELMAN, Senior Program Officer, Grand Challenge on Climate Change, Human Health, and Equity, National Academy of Medicine
BRENT HEARD, Program Officer, BEES
KASIA KORNECKI, Program Officer, BEES

ALEX REICH, Program Officer, Board on Atmospheric Sciences and Climate
CATHERINE WISE, Program Officer, BEES
HANNAH STEWART, Associate Program Officer, Board on Environmental Change and Society
REBECCA DeBOER, Research Associate, BEES
KYRA HOWE, Research Assistant, BEES
JASMINE VICTORIA BRYANT, Program Assistant, BEES
KAIA RUSSELL, Program Assistant, BEES

BOARD ON ENERGY AND ENVIRONMENTAL SYSTEMS

BOARD ON ATMOSPHERIC SCIENCES AND CLIMATE

BOARD ON ENVIRONMENTAL CHANGE AND SOCIETY

Reviewers

This Proceedings of a Workshop was reviewed in draft form by individuals chosen for their diverse perspectives and technical expertise. The purpose of this independent review is to provide candid and critical comments that will assist the National Academies of Sciences, Engineering, and Medicine in making each published proceedings as sound as possible and to ensure that it meets the institutional standards for quality, objectivity, evidence, and responsiveness to the charge. The review comments and draft manuscript remain confidential to protect the integrity of the process.

We thank the following individuals for their review of this proceedings:

CHELSEA FOWLER, National Academies of Sciences, Engineering, and Medicine
KATHRYN J. JACKSON (NAE), Independent Consultant
CAROL LEWIS, Texas Southern University

Although the reviewers listed above provided many constructive comments and suggestions, they were not asked to endorse the content of the proceedings nor did they see the final draft before its release. The review of this proceedings was overseen by CHERRY MURRAY (NAS/NAE). She was responsible for making certain that an independent examination of this proceedings was carried out in accordance with standards of the National Academies and that all review comments were carefully considered. Responsibility for the final content rests entirely with the rapporteur and the National Academies.

Contents

Overview

Discussions of reducing CO_2 emissions frequently focus on technologies. While technologies are clearly instrumental in transitioning away from fossil fuel–based energy and toward a decarbonized economy, decisions about which technologies are prioritized, how they are implemented, and the policies that drive these changes will have profound effects on people and communities, with important implications for equity, jobs, environmental and energy justice, health, and more. The National Academies of Sciences, Engineering, and Medicine's Committee on Accelerating Decarbonization in the United States: Technology, Policy, and Societal Dimensions was tasked with assessing the broad range of technological, policy, and societal dimensions of decarbonizing the U.S. economy. The committee's 2021 report[1] provided the U.S. government with a roadmap of equitable and robust decarbonization policies. The committee's next report will address the broader range of policy actors who play a role in equitable energy transition.

To inform its deliberations, the committee hosted a 1-day workshop to discuss critical issues of equity and justice during the energy transition. The goal of the workshop, titled Pathways to an Equitable and Just Transition: Principles, Best Practices, and Inclusive Stakeholder Engagement, was to move beyond energy technologies and elicit ideas and insights to inform the development of principles, best practices, and

[1] National Academies of Sciences, Engineering, and Medicine, 2021, *Accelerating Decarbonization of the U.S. Energy System*, Washington, DC: The National Academies Press, https://doi.org/10.17226/25932.

actionable recommendations for a broad range of policy actors and stakeholders in order to fully operationalize equity, justice, and inclusion. The workshop was held in person in Washington, DC, on July 26, 2022, with additional speakers and attendees joining virtually via Zoom. Recordings, slides, and other materials from the workshop are available online.[2]

Setting the stage for the workshop discussions, committee member Michael Méndez, University of California, Irvine, noted that a large portion of CO_2 emissions come from distributed sources such as personal vehicles, home heating and air conditioning, and other energy-consuming equipment that individuals own or rent. This means that action on the part of individuals will play a crucial role in decarbonizing the economy, which has significant energy and environmental justice implications. "Navigating through the transformation of that infrastructure through energy efficiency, adoption of new technologies, and the switch to clean electricity—and ensuring that everyone is able to participate and derive significant benefits as a result—will require the active participation, support, and engagement of the American public," said Méndez. "It is crucial, therefore, that we fully understand that challenge."

The workshop was organized into four sessions focusing on the following topics: public health, safety, and community resilience; jobs and workforce development opportunities; access to transition technologies, infrastructure, and programs; and energy affordability and burdens. Following presentations and brief discussion sessions on each of these topics, attendees from academia, government agencies, nonprofits, and community-based groups joined roundtable discussions to reflect further on these topics and identify key needs and ideas. The workshop concluded with a wide-ranging discussion of challenges and opportunities involved in setting the United States on the path toward a just and equitable energy transition.

Several key themes and needs emerged over the course of the discussions. Expanding on the workshop's overall context, many speakers stressed that technology alone cannot solve the problem of decarbonization. Energy access and use is personal and local, and energy policy affects people's health, finances, and quality of life. Disparities in access to clean energy technologies, energy affordability, and energy insecurity interact with and often reinforce disparities in other areas of society, such as poverty, education, worker exploitation, access to health care, and access to a healthy environment.

---

[2] National Academies of Sciences, Engineering, and Medicine, 2022, "Pathways to an Equitable and Just Transition Workshop," https://www.nationalacademies.org/event/07-26-2022/accelerating-decarbonization-in-the-united-states-technology-policy-and-societal-dimensions-pathways-to-an-equitable-and-just-transition-workshop.

Despite the many challenges of advancing a clean energy transition both swiftly and thoughtfully, recent research and city- and state-level initiatives offer valuable insights and lessons learned to inform future efforts to create a just energy transition in which all communities share the benefits and burdens. Participants underscored the importance of data to support a shared understanding of energy burdens and energy poverty. Metrics on energy costs, reliability, usage, and needs—across different types of households and communities—will be vital to designing solutions that are appropriate for the context of people's lives and guiding where and how assistance is most needed.

Decarbonization will undoubtedly have implications for people employed in fossil fuel–based industries that will be phased out, pointing to a critical need to consider ways to make the energy transition just and equitable for workers in addition to energy consumers and society more broadly. Participants explored how decarbonization can be leveraged as an opportunity to create high-quality, long-term jobs and careers and examined roles for government programs and grassroots organizations in ensuring historically disadvantaged communities can realize the benefits of this opportunity. Participants also stressed the importance of supporting resilience and foregrounding environmental justice and health for both workers and their communities.

Across all of these areas, participants emphasized the central role of broad-based coalitions to create community-driven and co-designed solutions. This means taking a people-first approach, being attentive to community priorities, and seeking locally relevant solutions. Participants also emphasized the importance of learning from local, small-scale projects that could have a larger impact and the authentic inclusion of and shared decision-making with those whose work, health, or communities have been directly affected by fossil fuels.

Equity is foundational to ensuring that the clean energy transition is truly transformative; communities do not benefit from clean energy technologies when they cannot access them. Given that funding many of the changes discussed as part of decarbonization could necessitate rate increases for customers, there is a risk that low-income and historically disadvantaged groups could become further overburdened if the transition is not thoughtfully managed and centered in the lived experiences of these communities. Participants stressed the need to ensure that those who face the greatest burdens from the fossil fuel economy—and those who face the highest rates of energy insecurity today—are "first in line" to benefit from the clean energy transition, rather than being left behind.

1

Public Health, Safety, and Community Resilience

We have a fair body of evidence showing that the transformations we're talking about will, for the workers and communities, have huge health and economic benefits.
—Linda Rudolph, Public Health Institute[1]

One of the workshop's first simultaneous sessions focused on the interplay between climate change and community health. Michael Méndez, University of California, Irvine, introduced panelists Garvin Heath, National Renewable Energy Laboratory (NREL); Linda Rudolph, Public Health Institute; and Jason Beckfield, Harvard University. Speakers offered opening remarks before engaging in a wide-ranging discussion of the challenges and opportunities involved in incorporating public health perspectives in decarbonization efforts, and noted the following:

- Studies have demonstrated that it is possible for decarbonization measures to simultaneously reduce emissions, improve resilience, improve public health, and reduce environmental justice disparities.
- Achieving a just transformation of communities, people, work, industry, and government requires an emphasis on equitable benefits for workers and for communities who bear the greatest climate burdens.

[1] Linda Rudolph, Public Health Institute, presentation to the workshop on July 26, 2022.

- People-first approaches that are attentive to community priorities and seek locally relevant solutions can help to elevate health as a key component of energy transitions and inform impactful and equitable decarbonization strategies.

LESSONS FROM LA100

The city of Los Angeles, California, has embarked on an ambitious plan to transition the city to 100 percent renewable electricity by 2045. Heath discussed LA100,[2] a study conducted by NREL in partnership with the Los Angeles Department of Water and Power (LADWP) to assess the pathways toward achieving that goal. For the study, researchers evaluated the technical feasibility of various approaches as well as their implications for public health and environmental justice. Despite significant challenges, including the geographic constraints imposed by the city's position between mountains and ocean and its high current pollution levels, the study identified pathways that could result in significant gains for air quality through the electrification of the power sector and its loads from light-duty vehicles, buses, residential and commercial buildings, and ports, among other measures. "We demonstrated in this study how a global megacity could achieve aggressive decarbonization goals while also achieving billions of dollars of environmental health co-benefits and reducing environmental justice disparities," said Heath. One challenge, Heath noted, was that researchers sought to predict future environmental justice impacts, but existing empirical tools are largely designed to assess environmental justice impacts retrospectively, suggesting a need for better prospective environmental justice assessment tools.

FOCUSING ON WORKERS

Rudolph described working many years ago with workers in the oil industry who face negative impacts from occupational exposures and from energy transitions. She stressed that the loss of jobs and industries in a community can itself have adverse health impacts. "We all know that economic well-being and security is a truly fundamental determinant of health and a precursor to health and well-being," said Rudolph. In addition, even "clean" energy technologies do not necessarily equate to clean and healthy working conditions. Some green economy jobs can be just as dangerous or low-wage as fossil fuel jobs, especially overseas where workers who supply necessary raw materials for products such

[2] National Renewable Energy Laboratory, 2021, "LA100: The Los Angeles 100% Renewable Energy Study," March, https://maps.nrel.gov/la100/la100-study/report.

as electric-vehicle batteries often work in harsh conditions. Therefore, she emphasized that creating a just transition will require collaboration among labor, civil rights, and environmental justice movements to ensure that workers have safeguards that prioritize climate, health, and equity, and empowerment. Achieving this level of collaboration requires meaningful community and worker engagement in decision-making to link the energy transition with a transition to a more equitable society.

ADDRESSING SOCIAL BARRIERS TO DECARBONIZATION

While many people tend to focus on technological or financial barriers to decarbonization, Beckfield argued that social barriers are often more important. Through his work studying decarbonization on the U.S. Gulf Coast, he found that factors such as environmental racism, a sense of resentment between urban and rural areas, political polarization, and the feeling of disrespect experienced in many communities can drive distrust, disengagement, and stasis. As a result of income inequality and other forms of inequity, many communities lack social cohesion or a sense of shared fate.

To attend to these issues in the quest to accelerate decarbonization, Beckfield offered several suggestions. First, he said it is important to gain a pragmatic understanding of how people define their own problems and priorities. To make solutions more relevant to local communities, he suggested focusing on regionalized social solutions and undertaking an ethnographic study to localize national projects such as those being pursued at national laboratories. He added that cohort-specific labor market policies could increase overall effectiveness of labor policies by attending to the different needs of workers in different age groups. To strengthen research in this area, he suggested undertaking a comprehensive study of how people relate to energy systems, accounting for cumulative health impacts from the energy transition, and bringing the impacts and implications of political conflict and unequal power resources into the analytical frame, rather than treating it as separate. Finally, he suggested focusing on scenarios that do not involve a national carbon price, which he asserted is unlikely to be adopted.

DISCUSSION

Building on the speakers' comments, Nancy Krieger, Harvard University, served as discussant for a panel discussion examining approaches to better understand the potential health benefits and harms of the coming energy transition. To open the discussion, Krieger posited that rather than a "transition," decarbonization will be a "transformation" that will revolutionize global relationships, especially between people and work, and create health benefits not only for a person's entire life span but also for future generations. Success will require learning from past projects, engaging people, understanding their problems, and including their input in the transformation. To accomplish this, she said that it is crucial for scientists to look beyond their own perspectives and work with many other groups with a vested interest in creating a healthy, equitable, and just society.

Thinking on Different Scales

Panelists discussed the different scales at which the social dimensions of decarbonization may be understood. Noting his agreement that "transformation" is a helpful framework that encapsulates how radical the impacts of decarbonization will be, Beckfield said that the geographic scale is critical, because the U.S. energy system is heterogeneous, and different parts of the country have very different needs.

Giving greater emphasis to the health impacts of decarbonization can help to put cost scales into perspective. Heath said that transition costs often seem large, but monetizing accrued health benefits would expand the definition of cost (and benefit) to include the expected social and health improvements for current and future generations. Rudolph added that these improvements are already well documented. If done right, she said, "we have a fair body of evidence showing that the transformations we're talking about will, for the workers and communities, have huge health and economic benefits."

As for the time scale, Rudolph stated that government is generally very slow to act, but the transformational policy and systems changes under discussion must happen at a radically faster pace. She suggested that highlighting beneficial health and equity impacts, perhaps by adding formal requirements for comprehensive health and equity analyses during decision-making, could help to speed action. She added that those analyses should emphasize benefits and risks to workers and communities whose health and economic well-being have often been ignored by the fossil fuel industry.

Carlos Martín, Brookings Institution and Harvard University, asked how to balance the tension between the need to move quickly to mitigate future climate effects and rectify past injustices, and the time it takes to conduct responsible long-term assessment, modeling, and public engagement. Rudolph acknowledged that it is challenging to pursue studies, community engagement, and decision-making simultaneously, but it is essentially the only option we have. Fortunately, more resources are being developed to help guide effective community-engaged efforts. Building on this point, Beckfield suggested that scientists could study models from other countries that have undergone energy transitions, draw from the work of the White House Environmental Justice Advisory Council, and adapt models for community engagement from other disciplines, such as Louisiana's Strategic Adaptations for Future Environments (LA SAFE).[3,4]

Accounting for Health Benefits

Prompted by a question from Méndez, speakers discussed the importance of including more wide-ranging health benefit analyses in decarbonization planning, along with approaches for achieving this. Rudolph stressed that climate change is a health emergency and that community health must be considered as an integral part of the solution; including health benefit analyses in decarbonization planning is critical to identifying strategies that can simultaneously reduce emissions, improve health, and advance environmental justice. On the flip side, failing to adequately account for health in policymaking could deprioritize solutions with health co-benefits, leading to missed opportunities, and even bring unforeseen health risks, especially to communities that already bear significant environmental and health burdens from climate change. For example, she noted that transportation transitions have mostly focused on switching personal vehicles from gas to electric, but buses and heavy-duty vehicles[5] may be more pertinent considerations in many climate-impacted communities. Heath added that the LA100 effort identified the electrification of light-duty vehicles and buses as critical to decarbonization. Rudolph also cautioned that there has been little analysis of potential health risks from carbon capture and sequestration technologies, which she said also have the disadvantage of being expensive, unproven, and potentially likely to perpetuate the use of carbon-based fuels.

Heath agreed that integrating and quantifying health benefits within energy transition studies is critical, although he acknowledged that doing so adds to the challenges of study design. He suggested that more

[3] Environmental Protection Agency, 2022, "White House Environmental Justice Advisory Council," https://www.epa.gov/environmentaljustice/white-house-environmental-justice-advisory-council.

[4] Foundation for Louisiana, 2019, "Louisiana's Strategic Adaptations for Future Environments (LA SAFE)," https://lasafe.la.gov.

[5] Light-, medium-, and heavy-duty are classifications used to distinguish between different types of vehicles, generally based on size, weight, and purpose. Light-duty covers passenger cars, sport utility vehicles, and most pick-up trucks; medium- and heavy-duty covers larger vehicles such as delivery vans, buses, and tractor trailers. For more information, see Federal Highway Administration, 2020, "Figure 21: Law Enforcement Vehicle Identification Guide," U.S. Department of Transportation, https://ops.fhwa.dot.gov/publications/fhwahop10014/long_f21.htm.

accessible tools, such as air quality models with reduced complexity, will enable more scientists to quantify health benefits. Building on this point, Beckfield said that incorporating health benefits is challenging in a world that prefers single causes over simultaneous or long-term drivers. In addition, he said there is generally not enough data on the relationship between energy jobs and health, a gap that points to a clear research need. However, on the plus side, incorporating health benefits can lead to better public engagement than climate alone; health problems have motivated environmental justice communities to find novel solutions to toxic exposures and led to surprising collaborations with industry and policymakers. "Environmental justice communities are not passive receptacles of toxic pollution—there's a lot of creativity on the ground that I think could be incorporated," said Beckfield.

Krieger further emphasized the importance of collaboration. Connecting the dots between research, people, and multiple environmental exposures can lead to a more coordinated understanding of harms and benefits across life spans and generations. To do this, she suggested it will also be necessary to tackle disinformation, anti-regulatory sentiment, and the notion that we must always make a choice between jobs and health. Rudolph agreed and suggested that, with funding, the public health community could create a well-resourced media campaign that explicitly calls out the health risks of fossil fuels and the disinformation campaigns of the fossil fuel industry, akin to the model of anti-tobacco public health campaigns, and reframes the climate debate away from the idea that addressing environmental problems must come at the expense of jobs.

Given that the United States is such a heterogenous country with regard to geography, resources, and community composition, Clark Miller, Arizona State University, asked panelists to comment on which modeling and quantitative approaches to risk and benefits analyses might be most effective and accurate. Heath recommended agent-based modeling, which is a method within complex system science that can differentiate among communities using social and technical factors for individuals and organizations. From a different standpoint, Beckfield encouraged complementing modeling with a more social, people-first perspective. Narratives can sometimes be more valuable than modeling because they help identify the processes and motivations behind key decisions. Rudolph agreed, adding that narratives have brought insight to transitions beyond energy, such as how forests are managed or food is supplied. Scaling up inclusive conversations requires resources, but drawing out the key drivers from a people-first perspective can help to identify pathways to healthy transformations. "We don't have to do each analysis over in every city; we need to figure out how to take what we learn and move to the next point on our trajectory toward transforming these systems," said Rudolph.

2

Jobs and Workforce Development Opportunities

These transitions are local and hyper personal. They're going to require tailored support and programs unique to each situation, and that just takes time and effort and sustained resources to understand the context. Only then do I think that we'll be able to have a just and equitable transition.
—Sarah Truit, National Renewable Energy Laboratory[1]

The workshop's second simultaneous session focused on jobs and workforce development opportunities. Panelists included Mijin Cha, Occidental College; Madeline Janis, Jobs to Move America; and Sarah Truitt, NREL. Following panelists' opening remarks, Roxanne Johnson, BlueGreen Alliance, moderated an open discussion. Speakers noted the following:

- Job and workforce transitions are highly personal and highly local. National-level trends in the green economy play out unevenly across geographies and communities, opening opportunities for some while leaving others behind.
- Decarbonization is a market-shaping opportunity to create high-quality, long-term jobs and careers. Government programs and procurement policies can play a major role in ensuring

[1] Sarah Truit, National Renewable Energy Laboratory, presentation to the workshop on July 26, 2022.

historically disadvantaged communities can realize the benefits of this opportunity.

- A just workforce transition will benefit from strong government and grassroots support and resources, dedicated and sustained workforce development funding, and targeted efforts to achieve structural, procedural, distributional, and transgenerational equity.

CHALLENGES AND OPPORTUNITIES FOR A JUST TRANSITION

Cha discussed how the challenges encountered as part of the energy transition fit into the broader picture of workforce transitions more generally. Decarbonization represents an immense opportunity to create better jobs so that all workers can obtain high-quality, family-sustaining careers in their communities. However, Cha cautioned that this not going to happen on its own. Creating a just transition for workers will require addressing the root causes that have contributed to the unjust working conditions in many industries today and to unjust workforce transitions in the past.

With a federal minimum wage stuck at $7.25 per hour and generally weak worker protections, Cha argued that the United States has a poor track record of creating good jobs. Jobs in fossil fuel industries, such as coal mining, are considered good jobs in their communities—not because they are inherently good, but because of the history of worker struggle and unionization in these industries. This means that workers transitioning out of fossil fuel industries are often unable to find new jobs that offer a similar level of economic stability. "You're asking a coal miner to leave their job in coal mining, which is one of the last decent jobs in the region, for something like a Walmart job," Cha said. "That's not a just transition."

The quality of jobs matters, not just the quantity; people want not only jobs, but careers. One green economy opportunity that is often touted is solar installation; however, Cha noted these jobs are typically temporary because there is a limit to the amount of solar infrastructure that will be economical to install in any given region. However, a worker who trains to become an electrician who can repair and maintain solar installations will have more economic security than a temporary solar installer. This is just one example pointing to the broader need to tie job training to the supply and demand for workers, and also to think regionally when encouraging job training and the growth of green industries, Cha said. The geographical disconnect between where jobs are being lost and where jobs are being created can pose significant barriers for communities that are heavily dependent on fossil fuel–based industries.

To avoid repeating past patterns of exploitation, Cha outlined four pillars for a just transition: strong and sustained government support;

dedicated and long-term funding for workforce transition; economic diversification to reduce local dependency on one industry; and strong, diverse coalitions. She stressed that climate change is a political problem that needs a grassroots-based political solution. "The transition must come from the ground up," said Cha. "Coming from the ground up is important to ensure that we have appropriate policies, but also to ensure that we build the political power we need to have a just transition."

A FRAMEWORK FOR SHAPING MARKETS

Janis discussed how decarbonization represents a market-shaping opportunity to create high-quality jobs in communities that have suffered from the fossil fuel industry and carbon-intensive economy. She outlined a framework for market-shaping across three stages: planning and policy, contract and practice, and implementation and evaluation. In all of these stages, Janis said that an important role for government is to require equity and high-quality jobs for every contract and subsidy rewarded, taking cues from the Justice40 Initiative's requirement that 40 percent of federal energy investment flows to disadvantaged communities.[2]

In the planning and policy stage, Janis said it is important to include all stakeholders from the beginning, especially those who have been historically excluded and will benefit most from the energy transition. At this stage, rather than one-size-fits-all strategies, she said it is valuable to create policy and program "toolkits" tailored for each job type, such as construction, manufacturing, operations and maintenance, and recycling. In the contract and practice stage, Janis said that public funding (which accounts for the lion's share of green energy transition funding) and public procurements (which account for a sizeable portion of manufactured goods purchases) should include incentives for high-quality jobs and equity, as well as subsidies to target diversity and meet regional needs. In the implementation and evaluation phase, policies should be enforced, monitored, and adjusted to continue to enable communities to negotiate training, hiring, and other benefits. Janis said that this framework enabled her organization to enter robust community-benefiting partnerships with three responsible, equitable electric bus companies, which included targeted hiring, U.S.-based manufacturing, and unionization commitments.

[2] White House, 2022, "Justice40 Initiative," https://www.whitehouse.gov/environmentaljustice/justice40.

ATTENDING TO PERSONAL AND LOCAL PRIORITIES IN WORKFORCE TRANSITIONS

Building on the other speakers' comments, Truitt stressed that the tools and technologies for a green economy exist, but work is needed to ensure that the opportunities reach those who need them the most. At a national level, fossil fuel jobs are declining while jobs in renewable energy and other green industries are rising. However, the national statistics fail to capture the heavy burdens being felt in communities that have been highly dependent on fossil fuel industries, where job losses are both personal and local. For those who are losing jobs, the ability to transition to new industries involves a wide variety of factors, from geographic location to wages to corporate culture. At the community level, localities face uncertainties regarding the future of their tax base, net number of jobs, and programmatic priorities. In addition, there is much more work to do to diversify the clean energy workforce, where women and Black workers remain underrepresented.

There are resources to help employers and policymakers create and train workers for high-quality, family-sustaining, clean energy jobs across the United States, including for underrepresented groups.[3,4,5,6,7] While the Justice40 Initiative is an important driver, the government's role is not limited to this initiative. Truitt described the U.S. Department of Energy's Inclusive Energy Innovation Prize, which incentivizes community-based innovative pathways, including workforce development, to create a just and equitable energy transition.[8] In these and other efforts, Truitt said it is

---

[3] E.P. Louie and J.M. Pearce, 2016, "Retraining Investment for U.S. Transition from Coal to Solar Photovoltaic Employment," *Energy Economics* 57:295–302, https://doi.org/10.1016/j.eneco.2016.05.016.

[4] S. Truitt, J. Elsworth, J. Williams, et al., 2022, "State-Level Employment Projections for Four Clean Energy Technologies in 2025 and 2030," https://www.nrel.gov/docs/fy22osti/81486.pdf.

[5] A. Tomer, J.W. Kane, and C. George, 2021, *How Renewable Energy Jobs Can Uplift Fossil Fuel Communities and Remake Climate Politics*, Washington, DC: Brookings Institution, https://www.brookings.edu/research/how-renewable-energy-jobs-can-uplift-fossil-fuel-communities-and-remake-climate-politics.

[6] E2 (Environmental Entrepreneurs), American Council on Renewable Energy (ACORE), Clean Energy Leadership Institute (CELI), and BW Research, 2020, *Clean Jobs, Better Jobs: An Examination of Clean Energy Job Wages and Benefits*, Washington, DC: E2, https://e2.org/wp-content/uploads/2020/10/Clean-Jobs-Better-Jobs.-October-2020.-E2-ACORE-CELI.pdf.

[7] A. Dewey, J. Mah, and B. Howard, 2021, "Ready to Go: State and Local Efforts Advancing Energy Efficiency," Washington, DC: American Council for an Energy-Efficient Economy, https://www.aceee.org/toolkit/2021/11/ready-go-state-and-local-efforts-advancing-energy-efficiency.

[8] U.S. Department of Energy (DOE), 2021, "Inclusive Energy Innovation Prize," https://americanmadechallenges.org/challenges/inclusiveenergyinnovation/index.html.

critical to attend to structural, procedural, distributional, and transgenerational equity. "These transitions are local and hyper personal," said Truitt. "They're going to require tailored support and programs unique to each situation, and that just takes time and effort and sustained resources to understand the context. Only then do I think that we'll be able to have a just and equitable transition."

DISCUSSION

In an open discussion, panelists explored approaches to navigating challenges, expanding opportunities, and the need for larger reforms.

Navigating Challenges

Johnson asked each panelist to name the biggest challenge to a just transition from a workforce perspective. Cha replied that the biggest challenge in her view is a lack of proactive long-term planning. The transition has already begun, but many affected workers are simply being left behind. "Most of these coal communities have seen job loss, coal is already in the transition, and yet there are no supports for these communities or workers," said Cha.

Truitt named fragmentation of the clean energy industry as the biggest challenge and suggested that more effective cross-industry collaboration could help to identify transferable skills and facilitate career flexibility. Janis stated that the biggest challenge lies in countering the long-standing assumption that market solutions are the answer, a notion that in her view has led the government to fund ideas with little accountability. To counter this, her organization is focused on changing the outdated Uniform Guidance of the Office of Management and Budget[9] to require equitable policies and Justice40 principles at every level, including across the entire supply chain.

Expanding Opportunities

Committee Chair Stephen Pacala, Princeton University, asked panelists to comment on how to navigate conflict among workers competing for scarce jobs. Janis shared that unionized workers who maintained public diesel buses were first opposed to fleet electrification because they

[9] U.S. Federal Government Agencies, 2014, "Federal Awarding Agency Regulatory Implementation of Office of Management and Budget's Uniform Administrative Requirements, Cost Principles, and Audit Requirements for Federal Awards," *Federal Register* 79:75867, December 19, https://www.federalregister.gov/d/2014-28697.

feared job loss, so her organization included retraining requirements in the procurement contracts in order to transition workers. This model could work holistically across industries, and places the transition burden on companies, not workers. In addition, "second chance" programs to replace retirees or commitments to hire from underrepresented groups can expand the job pool to those who have been shut out from work altogether. Truitt agreed, noting that targeted programs to address specific challenges can be more successful than broader, less targeted workforce efforts.

Cha added that one reason people mourn the loss of fossil fuel jobs is that they are often much higher-paying, but often overlooked is the fact that the work is very dangerous. "It is important to remember that even fossil fuel workers are poisoned by the fossil fuel economy," Cha said. Rather than simply seeking to swap any one job with another, she said the focus should be on ensuring that there are enough good-paying jobs for those leaving fossil fuel work or those historically shut out from it *and* on ensuring that renewable energy jobs are safe and pay a living wage.

The Need for Larger Reforms

Lee Anderson, Utility Workers Union of America (UWUA), agreed that job loss was a serious challenge, noting that 20 percent of UWUA workers have lost their jobs since 2010 and have received no government support. He asked if it was possible to expand the political narrative from local jobs and communities to encompass wider labor, health care, education, and student loan reforms to truly improve people's lives.

Janis stated that while politics may not be amenable to wider reform right now, money and labor do hold power. Attaching good policies to every procurement dollar can force high-quality job creation and return power to the unions, especially in the South, where many clean energy companies are taking advantage of weak labor and environmental laws, she said. For example, her organization was able to win community benefits agreements for factory workers in Alabama and Kentucky.

Cha agreed that financial influence is important, adding that those who see decarbonization as part of the larger issues of resource extraction and worker exploitation are gaining momentum and could force the kind of structural change Anderson described. She suggested that policymakers and organizers should help people see how their immediate concerns, such as job loss or health care, are connected to the larger issues of inequality and the climate crisis.

3

Equitable Access to Transition Technologies, Infrastructure, and Programs

As more investments are allocated for energy-related projects, equity—as in sharing the benefits, as well as the burdens of policy outcomes—will become an important subject that we need to address in the coming years.
—Pjoy Chua, Los Angeles Department of Water and Power[1]

The third session of the workshop focused on equitable access to transition technologies, infrastructure, and programs. Paty Romero-Lankao, NREL, served as moderator. She opened the session by underscoring the importance of learning from past experience. Studying methods that have succeeded or failed during previous times of transition can help overcome challenges and identify opportunities for effective community involvement, program design, and just implementation. The session focused on uncovering what Romero-Lankao referred to as "the deep infrastructural elements of the transition: How do we reach out to communities? How do we meet them where they are? What kind of programs and policies need to be designed and implemented to really shape the way we do business?" Speakers Pjoy Chua, LADWP; Leslie Aguayo, Greenlining Institute; Darnell Grisby, Beneficial State Foundation; and James Wilcox, New York State Energy Research and Development Authority (NYSERDA) offered principles, case studies, and lessons learned relevant to ensuring equitable access in the energy transition, including the following:

[1] Pjoy Chua, Los Angeles Department of Water and Power, presentation to the workshop on July 26, 2022.

- Equity is foundational to ensuring that the clean energy transition is truly transformative. Communities do not benefit from clean energy technologies when they cannot access them.
- City and state-level initiatives are making progress and offering valuable lessons learned for efforts to create a just energy transition in which all communities share the benefits and burdens.[2]
- Listening to community voices, working to heal past injustices, and engaging communities with effective co-design approaches can inform solutions that are equitable and attentive to community needs and priorities.

ADDRESSING EQUITY IN TRANSITION PLANNING

As Heath discussed in the workshop's first session, the LA100 study provides a comprehensive understanding of the complex requirements, trade-offs, and benefits of different pathways to meet ambitious clean energy goals for the city of Los Angeles. The study concluded that it is possible to achieve 100 percent renewable energy even earlier than the target date of 2045 but that the effort will require significant investments (on the order of $50 billion to $80 billion) and job creation. However, it did not directly address equity as part of the analysis of pathways to achieving the city's target. Chua described how a follow-up study now under way, LA100 Equity Strategies, is addressing equity with a focus on ensuring the disadvantaged communities that make up 47 percent of the city's residents are not overburdened or left behind in the city's energy transition. By addressing questions around affordability, who benefits from the transition, and effects on frontline communities, the overarching goal of this effort is to chart a path for all Los Angeles communities to share the equitable distribution of the benefits and burdens of a clean energy transition.

In accordance with the effort's core principles, the project aims to create a process and outcomes that are inclusive, community-driven, collaborative, accessible, affordable, measurable, and meaningful, prioritizing equity outcomes such as reducing energy burdens, expanding access to clean energy jobs, increasing access to clean mobility, assessing impacts to housing, improving reliability and resiliency, and improving access to rooftop solar. Chua stressed the importance of acknowledging past injustices while shaping the future, underscoring the principle of energy justice, which refers to the goal of achieving equity in both the social and economic participation in the energy system while also remediating social, economic, and health burdens on those historically harmed by the energy system.

---

[2] National Renewable Energy Laboratory, 2022, "LA100: The Los Angeles 100% Renewable Energy Study and Equity Strategies," https://maps.nrel.gov/la100/equity-strategies#home.

Thoughtful governance structures with meaningful community participation, complemented by inclusive community meetings and listening sessions, will play a crucial role in building the community support needed to overcome challenges, such as securing funding through voter approval or rate adjustments. Chua suggested that the project can help to inform equitable energy transitions in other communities and at every level of government. "As more investments are allocated for energy-related projects, equity—as in sharing the benefits, as well as the burdens of policy outcomes—will become an important subject that we need to address in the coming years," Chua said.

OPERATIONALIZING EQUITY

Highlighting lessons learned from efforts to expand access to electric vehicle (EV) charging infrastructure, Aguayo outlined four key actions to advancing equitable access to clean energy: defining and operationalizing equity; bringing benefits to people, not places; removing barriers; and integrating climate resilience.

The Greenlining Institute defines equity as the practice of transforming behaviors, institutions, and systems that disproportionately harm people of color by increasing access to power, redistributing resources, and eliminating opportunity barriers. Aguayo stressed that equity is not an obstacle to progress—it is the foundation needed to meet climate goals and ensure that all communities have access to clean energy and a high-quality life. Operationalizing equity requires a multistep process that starts with defining equitable goals and ends with creating measurable progress.

Bringing benefits to people, not places, means targeting funding and investment for frontline communities and avoiding displacement for vulnerable residents. Without specific attention to the needs of these communities, they can be left out of the benefits of clean energy investments. For example, charging stations that are only available to employees of a particular office building do not actually help increase access in nearby communities, even though it may appear these residents are close to charging infrastructure when looking at a map.

Removing barriers is critical to helping communities of color recover from historical injustices (such as redlining), capitalize on new investments, and build intergenerational wealth, Aguayo said. For example, a California program, Green Raiteros, received funds from the California Air Resources Board to turn an informal ride-sharing service into an "indigenous Uber."[3]

---

[3] M.G. Ortiz-Briones, 2021, "LEAP's Green Raiteros Electric Vehicle Ridesharing Program Gets Loaners to Expand Services," *Fresno Bee*, September 16, https://www.fresnobee.com/vida-en-el-valle/noticias/california-es/fresno/article253432099.html#storylink=cpy.

Finally, Aguayo stressed that integrating climate resilience into siting energy infrastructure requires consideration not only of climate impacts and disaster scenarios, but also consideration of the services provisioned by and support needed for the broader transportation and energy infrastructure and workforce. Spending on pre-disaster mitigation has been proven to lower disaster recovery costs, Aguayo noted, pointing as an example to a California school district that was able to find creative uses for its fleet of electric school buses, unneeded when schools were closed due to the COVID-19 pandemic, to deliver food and Wi-Fi to those in need.[4]

A FOCUS ON FINANCE

Grisby described how the Beneficial State Foundation's Clean Vehicle Assistance Program (CVAP) is tackling energy transition equity by addressing financial barriers to EV ownership. With its own bank, the program offers low-interest loans to help people, especially low earners or those with bad credit, buy EVs. Because charging is also important for EV access, CVAP also offers grants for home chargers and credit for public chargers and is working to incentivize multi-family home builders to include chargers for new construction. An initial pilot program, in partnership with several community organizations, helped more than 4,500 households purchase EVs and demonstrated broad interest in EV adoption.

In the future, Grisby said the program aims to grow further by reaching more buyers, especially in disadvantaged communities and areas with low public transit access, offering personal finance coaching, and optimizing participation incentives for low-income communities. Finally, Grisby noted that while expanding EV ownership is one important way to create a more equitable energy transition, transportation equity must include a full panoply of clean transit options, including making it easier and safer for people in all communities to get around by walking or biking.

CO-DESIGNING SOLUTIONS

Wilcox described NYSERDA's efforts to build capacity for a just transition in the context of advancing the overall energy and climate plans for the state of New York. In 2019, New York State passed landmark climate legislation supporting a just, equitable, and inclusive transition to

[4] Federal Emergency Management Agency (FEMA), 2018, "Natural Hazard Mitigation Saves Interim Report," https://www.fema.gov/sites/default/files/2020-07/fema_mitsaves-factsheet_2018.pdf.

net-zero energy while distributing 40 percent of climate investments to disadvantaged communities.[5] Key elements of these efforts include outreach to empower every New Yorker, engagement with researchers and innovators, and a community-focused and intentional co-design process to capitalize on diverse expertise. Several community-based initiatives have already begun. For example, the Equity and Climate Transformation Research Program is a scientifically rigorous inquiry combining social sciences and humanities-based, interdisciplinary approaches with quantitative, traditional energy systems research to materially advance transformative, inclusive solutions in underserved or overburdened communities. In addition, the Disadvantaged Communities Barriers and Opportunities Report[6] supports and informs community co-design processes.

To facilitate an effective and equitable co-design process, NYSERDA is also establishing a system to ensure that engaged communities can be appropriately compensated for their contributions. "One of the key barriers [to engagement] is that the labor of community-based organizations is not often recognized as expertise and ability to be compensated in the co-design process," Wilcox said. "So, we've initiated a process to make it easy and streamlined to actually be able to compensate community-based organizations substantially for the expertise that they provide the state." Prioritizing innovative, community-driven approaches is grounded in partnerships between academia, communities, and government and often requires breaking through institutional silos and formalizing a vision for true collaboration.

DISCUSSION

Romero-Lankao noted that large-scale transitions are messy and must address past injustices, include community voices, overcome cultural and institutional resistance, and embrace non-technical solutions. A participant asked if it is possible to support supply-chain development in disadvantaged communities or other peripheral areas. Wilcox replied that NYSERDA has a mandate to "think upstream" when considering suppliers, and the collaborative, coalition-based nature of the organization's work makes it easier for them to engage, share dialogue, and create

[5] New York State Senate, 2019, "Climate Leadership and Community Protection Act," S.B. S6599, 2019–2020 Legislative Session, https://www.nysenate.gov/legislation/bills/2019/S6599.

[6] New York State Energy Research and Development Authority (NYSERDA), New York State Department of Environmental Conservation (NYSDEC), and New York Power Authority (NYPA), 2021, "New York State Disadvantaged Communities Barriers and Opportunities Report," Report Number 21-35, https://climate.ny.gov/-/media/Project/Climate/Files/21-35-NY-Disadvantaged-Communities-Barriers-and-Opportunities-Report.pdf.

solidarity with communities. Chua added that LA100 Equity Strategies is addressing equity by listening to community feedback to the report's recommended pathways, which include "Buy American" provisions and other means to bring equity to peripheral communities.

A participant asked Grisby and Aguayo what other clean energy areas their organizations are prioritizing. Grisby replied that while distributional impacts are often examined within a given sector or technology area, such as public transportation, more work is needed to understand and prioritize equity across technology areas. Aguayo agreed, noting that holistic, cross-sector solutions should start with a community needs assessment and seek technologies that offer multiple benefits, such as a community center's solar panels generating electricity for residents during a blackout. Chua added that studying previous investments and incentives with an equity lens can provide ideas for improving or expanding them to better reach disadvantaged communities.

4

Energy Affordability and Burdens

Technology deployment alone doesn't solve energy poverty, if people feel like when we actually deploy that technology, they can't afford to use it.
—Destenie Nock, Carnegie Mellon University[1]

Adrienne Hollis, National Wildlife Federation, introduced the workshop's fourth session, a discussion of existing energy affordability burdens and equitable measures to address them. The speakers were David Nahai, Nahai Consulting Services and LADWP (retired); Gregory Pierce, University of California, Los Angeles; Destenie Nock, Carnegie Mellon University; and Sacoby Wilson, University of Maryland. Speakers noted the following:

- Energy insecurity is a critical—yet often overlooked—issue in planning for an equitable energy transition.
- Energy poverty can be challenging to measure based on income level alone, suggesting a need for a multifaceted approach to identify vulnerable households and provide assistance where appropriate.
- Momentum is building for energy justice, the notion that those most adversely impacted by fossil fuel infrastructure should be "first in line" to benefit from a just transition.

[1] Destenie Nock, Carnegie Mellon University, presentation to the workshop on July 26, 2022.

AFFORDABILITY AND THE PATH TO 100 PERCENT RENEWABLES

Nahai highlighted LA100 as a model for other cities aiming to move away from fossil fuels, especially for communities that have not benefited from energy infrastructure in the past. As other speakers mentioned, LA100 and follow-on projects have elucidated pathways to achieve 100 percent renewable power with attention to safety, reliability, affordability, and equity. Local and statewide mandates for renewable energy have helped spur progress toward this ambitious goal, but achieving it will require substantial additional investment, significant updates to and deployment of renewable energy sources, grid modernization, higher levels of energy efficiency, and more support for EVs, as documented in LADWP's Strategic Long-Term Resource Plan.[2]

Nahai stressed that while rate increases may be inevitable in funding these investments, they must be structured equitably so as to avoid overburdening communities such as low-income customers and seniors living on a fixed income. Beyond energy rates, he added that energy efficiency programs, such as community solar projects, should specifically favor disadvantaged communities. Finally, Nahai said that a just transition must include workforce development programs that center both compensation and dignity. "If we're going to be a humane, civil, and decent society, we're going to have to take care of those who have worked honorably, who have given of themselves to make a better society for all of us," Nahai said.

A FOCUS ON ENERGY INSECURITY

Pierce underscored the critical role that energy affordability plays in a just energy transition. Renewable energy sources can raise energy bills and thereby increase burdens on customers, with particularly detrimental effects on lower-income households and households of color. These groups of customers already faced higher rates of energy insecurity before COVID-19 and saw substantial increases in shutoff and debt rates during the pandemic.[3] On the positive side, the pandemic created new momentum around energy affordability, with most states instituting shutoff bans and progressive payment policies. However, Pierce noted that most of these measures have now expired, allowing private and public utilities to pursue shutoffs and debt recovery focused on low-income households.

---

[2] Los Angeles Department of Water and Power, 2022, "Strategic Long-Term Resource Plan," https://www.ladwp.com/ladwp/faces/ladwp/aboutus/a-power/a-p-integratedresourceplanning.

[3] D. Hernández and J. Laird, 2022, "Surviving a Shut-Off: U.S. Households at Greatest Risk of Utility Disconnections and How They Cope," *American Behavioral Scientist* 66(7):856–880, https://doi.org/10.1177/00027642211013401.

To keep energy affordable, Pierce suggested that utility companies should offer progressive rates, limited fixed charges, discounts, and efficiency measures to reduce demand.[4] In addition, he suggested that utility debt can be resolved by both deepening existing assistance programs and creating new ones, perhaps modeled on Southern California Edison's Arrearage Management Program, a debt forgiveness payment plan.[5]

ADDRESSING HIDDEN FORMS OF ENERGY POVERTY

Nock discussed the importance of recognizing and addressing hidden forms of energy poverty as part of energy transition planning and evaluation. The concept of energy burden focuses primarily on the cost of energy, while energy insecurity encompasses cost as well as reliability and the risk of outages. Energy poverty encompasses both energy burden and energy insecurity, as well as supply concerns. Nock posited that no single metric captures energy poverty, but approaching the question from multiple angles can help identify households that are energy poor and guide approaches to ensure that they are not left behind in the energy transition. Affordability is key. "Technology deployment alone doesn't solve energy poverty, if people feel like when we actually deploy that technology, they can't afford to use it," said Nock. It is also important to acknowledge that while energy investments happen on the macroscale, many solutions need to be implemented at the microscale.

Using large-scale smart meter data assessments has helped Nock and her team to identify why some households under-consume energy and examine different metrics for understanding energy poverty. One approach is to look at energy bills as a percentage of household income. However, looking at energy costs only as a percentage of gross income often misses the full picture, because energy is one of a plethora of often non-negotiable expenses such as housing, student debt payments, commuting and vehicle costs, and groceries. Once these factors are accounted for, the full burden of energy as a percentage of disposable income is quite high in many low-income households.

Another approach is to look at how people perceive and manage their energy use. For example, asking people whether they actively limit their energy use can help to identify which households feel overburdened. In addition, tracking the outdoor temperature at which people turn on

---

[4] M. Graff, S. Carley, D.M. Konisky, and T. Memmott, 2021, "Which Households Are Energy Insecure? An Empirical Analysis of Race, Housing Conditions, and Energy Burdens in the United States," *Energy Research & Social Science* 79:102144, https://doi.org/10.1016/j.erss.2021.102144.

[5] Southern California Edison, 2022, "Arrearage Management Plan," https://www.sce.com/residential/assistance/Arrearage-Management-Plan.

their air conditioning reveals clear patterns. Nock and colleagues found that low-income households tend to wait to turn on air conditioning until the outdoor temperature is 4–7° hotter than higher-income households do.[6] This gap widens as energy prices increase, and this difference can have real impacts on health as many people die during heat waves despite having air conditioning at home.[7] Understanding the drivers and manifestations of energy poverty will be important considerations as utilities plan for future energy demands and governments consider the best ways to provide assistance to create an equitable energy transition.

ADVANCING ENERGY JUSTICE

Wilson outlined four pillars of energy justice: healthy and sustainable energy production; clean energy infrastructure, especially for those historically denied it; affordable energy; and uninterrupted service.[8] Noting that the burdens of fossil fuel infrastructure, extraction, emissions, transport, processing, combustion, and waste fall disproportionately on disadvantaged communities and communities of color,[9,10] he said that energy justice recognizes the right of those most impacted to be involved in the energy transition to ensure procedural equity, while distributional justice requires that they also benefit from the transition. "To think about energy justice, we need to move away from dirty energy infrastructure to clean energy infrastructure, and those who are disproportionately impacted by the legacy and the history and the current dirty energy infrastructure, they should be at the front of the line to get the clean energy infrastructure, and a higher proportion of it," said Wilson.

Energy justice also means going beyond addressing job loss to encompass energy sovereignty, freeing communities from hosting polluting energy infrastructure without benefitting from it, which he termed "environmental slavery." "People can produce their own energy and be free

[6] S. Cong, D. Nock, Y.L. Qiu, and B. Xing, 2022, "Unveiling Hidden Energy Poverty Using the Energy Equity Gap," *Nature Communications* 13:2456, https://doi.org/10.1038/s41467-022-30146-5.

[7] S.A. Iverson, A. Gettel, C.P. Bezold, et al., 2020, "Heat-Associated Mortality in a Hot Climate: Maricopa County, Arizona, 2006–2016," *Public Health Reports* 135(5):631–639, https://doi.org/10.1177/0033354920938006.

[8] R.J. Heffron and D. McCauley, 2017, "The Concept of Energy Justice across the Disciplines," *Energy Policy* 105:658–667, https://doi.org/10.1016/j.enpol.2017.03.018.

[9] National Association for the Advancement of Colored People (NAACP) Clean Air Task Force, 2017, "Fumes Across the Fence-Line: The Health Impacts of Air Pollution from Oil & Gas Facilities on African American Communities," https://naacp.org/resources/fumes-across-fence-line-health-impacts-air-pollution-oil-gas-facilities-african-american.

[10] NAACP, 2016, "Coal Blooded: Putting Profits Before People," https://naacp.org/resources/coal-blooded-putting-profits-people.

of the utilities and free of these systems which lead to these differential burdens," said Wilson.

Momentum is building, from the Justice40 Initiative, from Maryland's Greenhouse Gas Reduction Act,[11] and from other laws and programs, to ensure that historically impacted communities have a voice in energy investments and benefits. Wilson pointed to examples in which communities have rejected projects that did not promote energy justice; communities are also increasingly taking a lead role in conducting research to better understand health impacts. Looking ahead, Wilson suggested prioritizing energy justice in policy initiatives in several key areas, including port electrification; climate mitigation and resilience; community solar projects; access to and incentives for EVs, chargers, and green infrastructure; and equitable workforce development—as well as funding mechanisms to sustainably support these efforts.

DISCUSSION

In an open discussion, participants delved further into issues around energy equity and explored different approaches to reducing energy burdens for low-income households.

Cody TwoBears, Standing Rock Sioux tribal member and co-founder of the Indigenized Energy Initiative (IEI), highlighted particular energy challenges many Indigenous communities face. He noted that energy costs on reservations can be double the costs paid in nearby communities on a per kilowatt hour basis, and added that a reliance on poorly constructed, energy-inefficient houses further increases energy burdens in many communities. He asked what lessons or approaches the IEI might glean from other efforts to guide its battle for equitable energy and health policies. Nahai suggested leveraging the Justice40 Initiative to address building needed energy infrastructure, and creating energy co-ops on tribal lands to encourage energy sovereignty.

In reply to another participant's question, Pierce described several different models for lowering energy bills for resource-constrained customers. One model is percentage-of-income payments, an approach that has been deployed or piloted in several U.S. states, in which energy bills are capped at a particular percentage of income and utilities or other entities make up the difference. However, a downside of this is that it requires income verification and is therefore administratively burdensome, creating barriers to enrollment on the part of both utilities and customers. Approaches that are potentially less burdensome include

[11] Greenhouse Gas Reduction Act, 2021, "Maryland General Assembly," http://mde.maryland.gov/GGRA.

upstream rate design structures, efficiency measures, solar deployment, and crisis relief.

Nock added that demand response capabilities could be better leveraged to improve energy equity by not only flattening the peak part of the demand curve (essentially making higher-income customers a little less comfortable during peak times) but raising up the valleys at the low part of the demand curve (helping lower-income customers be more comfortable during lower demand times). This could be accomplished, for example, with smart meters and bill caps for lower-income customers, and it would have the added benefit of enhancing reliability by reducing the need to ramp production up as quickly during transitions from low to high demand periods. Miller asked for additional details on how equity metrics could be incorporated into policy and implementation. Nock replied that it is often difficult to incorporate those metrics because customer usage data are frequently privately controlled and demographics data are not always available. It is also hard to know what other expenses customers face for their basic needs that should be accounted for in determining energy burden in relation to income.

Jonathan Patz, University of Wisconsin–Madison, asked if communities had any legal recourse if utilities decide to reopen decommissioned coal power plants or delay planned closures. Wilson replied that the judicial history and recent rulings suggest this would be challenging, although the National Environmental Policy Act may provide some options. He suggested that communities should approach such events on a case-by-case basis, noting that such events would further delay the energy transition and impose greater environmental burdens on affected communities.

5

Discussion of Pathways for the Energy Transition

People made policies—and people can also change those policies.
—McKenna Dunbar, Sierra Club[1]

For the workshop's final session, participants gathered for a structured discussion of pathways for the energy transition. Groups were divided by general topic areas aligned with the sessions earlier in the day: public health and resilience, jobs and workforce, equitable access to decarbonization interventions, and energy affordability and burden. These sessions allowed for participants to dig deeper into topics raised in the prior presentations and panel discussions, as well as raise useful examples and case studies that could inform the committee's consensus recommendations (to release in mid-2023). Attendees first reported key take-aways from small breakout group discussions and then further expanded on their ideas and suggestions in a full group discussion.

SMALL GROUP DISCUSSION REPORT-OUTS

Committee Chair Stephen Pacala, Princeton University, noted that the committee has been charged with suggesting and implementing large-scale policies to help the United States achieve a fair and just transition. While there have been several successful small-scale projects tailored to specific communities, the challenge now is to create a national effort

[1] McKenna Dunbar, Sierra Club, presentation to the workshop on July 26, 2022.

to make a broader impact where it is most needed. To help inform the committee's thinking on future directions, he invited spokespeople from breakout groups to summarize the outcomes of their discussions, which focused on each of the four main workshop topics.

Public Health, Safety, and Community Resilience

Jonathan Patz, University of Wisconsin–Madison; Cynthia Harris, Environmental Law Institute; Leo Woodbury, Just Solutions Collective; McKenna Dunbar, Sierra Club; and Jenny Bradford, University of Wisconsin–Madison, summarized outcomes from their breakout group discussion focusing on public health, safety, and community resilience.

Many participants stressed the need to replace siloed approaches to health and energy with a broader, coalition-based approach in which groups share power, funding, and expertise. For governments, researchers, and other entities, this points to a need to establish meaningful partnerships and shared leadership with communities, whose rich expertise and actual needs are often overlooked. To do this effectively, participants underscored the importance of using decolonized and non-paternalistic language and advancing meaningful, intersectional, and substantive education and dialogue to match communities, agencies, and policymakers with foundations or programs that can offer resources and support. Finding pathways to identify, learn from, and scale up success stories can also help in these efforts.

Participants also pointed to a need for richer data to inform policy. This includes data to calculate true energy costs and burdens, better understand the nuances of energy use and needs in various communities, and examine the near- and long-term costs and health benefits of clean energy technologies across the entire life cycle, from equitable mining to generational impacts on public health.

Participants offered several concrete ideas for advancing these goals. First, some participants suggested program managers or bureaucrats should be required to liaise with community members to better understand their perspectives. For example, representatives from federal agencies could tour Southern communities to see firsthand the negative health and environmental impacts biomass wood pellet industry plants have had.[2] "Someone needs to get down on the ground [level] where they can see the actual problems that are going on and how communities want to address them," Woodbury summarized. Second, participants suggested focusing on supporting the innovative solutions that emerge

[2] Biomass pelleting plants process wood, agro-waste, and forestry residues like stalks and sawdust into pellets that can then be served as a substitute for fossil fuel combustion.

from community-based organizations, such as a program focused on installing water generators in churches to supply safe drinking water. As part of coalition-based efforts, governments and other entities can help to amplify the work of such groups, provide them with opportunities, and remove barriers such as funding and paperwork. Third, participants suggested that schools could better integrate advocacy and environmental stewardship into their curriculums, work to address climate anxiety among younger generations, and fund scholarships for future leaders.

Finally, participants suggested publicly addressing systemic negative practices through holistic conversations about equity, mental health, and environmental injustices at every age and policy level. By elevating community perspectives and acknowledging multiple viewpoints, these conversations can shape business and procurement practices, policies, technologies, and outcomes. "People made policies—and people can also change those policies," noted Dunbar. Participants suggested that such conversations are most impactful when they intentionally center health, humility, and community engagement while addressing systemic inequalities and underserved communities through a change in day-to-day relationships that avoid exploitative practices and give underserved communities appropriate power and sovereignty to determine their own fate.

Jobs and Workforce Development Opportunities

Roxanne Johnson, BlueGreen Alliance; Rebecca DeBoer, National Academies of Sciences, Engineering, and Medicine; and Paula Glover, Alliance to Save Energy, summarized their breakout group discussion on jobs and workforce development opportunities.

From the perspective of workforce development, participants stressed the need for retraining to create opportunities for long-term career transitions; a commitment to fostering diversity; and improved worker support and protections, including through unionization, wage and work standards, unemployment insurance, and strong retirement and relocation packages. To inform efforts to accomplish this, participants suggested a need for a central, cohesive entity to house and share best practices and existing tools; grassroots-led local and regional entities to bridge legislative gaps and create connected, broad-based coalitions; and structured and meaningful stakeholder engagement, input, and shared decision-making processes that incorporate stakeholder feedback, including a focus on renters and disadvantaged, fenceline, and Indigenous communities. Participants also noted the importance of removing or adjusting policies or industry subsidies that directly conflict with just transition goals.

Participants suggested several specific strategies for increasing job opportunities during the energy transition. For example, participants pointed to opportunities to create clean energy jobs around energy efficiency enhancements and climate mitigation and resilience strategies, such as weatherization and fuel-switching efforts. Participants also highlighted the value of creating interdisciplinary teams to study workforce outcomes and suggested partnering with K–12 schools to support a strong workforce pipeline.

Although speed is important, several participants underscored that "just" is the priority. Considerations for a just transition are often interconnected with broader issues, such as unionization, health care, and education. As such, participants underscored the importance of ending the exploitation of workers and vulnerable communities as part of a just decarbonization transition.

Equitable Access to Transition Technologies, Infrastructure, and Programs

Jason Beckfield, Harvard University, summarized three main points emerging from the discussions the breakout group on equitable access to transition technologies, infrastructure, and programs. First, it is critically important to think more about what is being transitioned *to*, and less about what the transition is *away from*. The answer will be different depending on the region, but solutions have to both scale up to be applicable to larger areas, and scale down to be relevant, helpful, and even joyful in the context of people's lives. Looking forward can inspire visions that are creative, affirmative, concrete, and local.

Second, participants stressed the importance of focusing on people's lived experiences, specific barriers, and technological needs by asking them what problems they are trying to solve and what they need in order to solve them. For example, Beckfield said, "When you're carrying a heavy load in your life already, you can't focus on doing an EV loan. This is very insightful, I think, and goes to the point that oftentimes community members don't have really the capacity to be aware of how they might benefit from different programs, or they may face very different kinds of barriers." Solving real problems requires embracing the ups and downs of trial-and-error, helping community organizations build capacity, creating financial innovations and targeted needs-based programs that avoid stigma, and helping with new technology maintenance. All of this in turn will also increase trust.

Third, politics is important and cannot simply be ignored. Navigating different communities' priorities can be polarizing, especially when an issue is framed around morality or notions about which groups or people are most "deserving" of solutions or assistance.

Energy Affordability and Burdens

Brent Heard, National Academies, summarized the breakout group discussion focusing on energy affordability and burdens. One key need the group identified is that both public and private utilities should cultivate a culture of embracing meaningful change. As part of this, participants suggested utilities should strive for more authentic engagement with communities and customers and better communication with customers about usage and assistance. The group also identified key data needs, including more refined air quality and location data, better measurements for a household's true energy burden, and a better understanding of how much energy people would consume in a more just world. Participants also suggested it would be valuable to analyze equity implications, accessibility, and quality-of-life outcomes as part of the planning and rollout of particular technology deployments.

Participants in this group posited that utilities and governments could combine assistance programs to reduce administrative burdens. For example, households qualifying for Supplemental Nutrition Assistance Program benefits could automatically be enrolled in utility assistance programs. Similarly, if someone has a health diagnosis that requires electrical medical equipment, an assigned case manager, perhaps working out of a local resilience hub, could file for utility shutoff protections or connect them to a community microgrid.

FULL GROUP DISCUSSION

In a final open discussion, committee members, speakers, and workshop participants explored the connection between the energy transition and other social issues and suggested several ideas for future learning opportunities.

Energy Transition and Social Issues

Pacala asked about the relationships between the energy transition and other social issues. Johnson stated that decarbonization and other problems are fully intertwined, and in order to gain support for the energy transition, people first need to feel secure in their communities and their lives. Romero-Lankao agreed, noting that energy is embedded in every aspect of society. In Los Angeles, for example, the city must account for renters, who make up 62 percent of the population, in any effort to increase building electrification. Krieger also agreed, adding that people engage with multiple ecosystems and infrastructures simultaneously—rather than segregating different aspects of their lives into different sectors—and successful discussions and solutions must acknowledge that fact.

Khalil Shahyd, National Resources Defense Council, also emphasized that climate change and social issues are fully intertwined and cautioned that people can be put at risk when arguments for expediency are used to argue against social justice. Most nations must simultaneously address climate change and social issues, but this requires deep conversation and cooperation—not only technical solutions. However, Shahyd suggested that the current political climate in the United States makes this level of cooperation nearly impossible. The decarbonization paths the United States takes can influence other nations, and Shahyd cautioned that if the United States emphasizes technology over social issues, the ripple effects could be devastating to the entire world's access to transportation, housing, and food.

On the other hand, Beckfield suggested that tying decarbonization to other political issues also has its downsides. He suggested that a just transition could be achieved faster and more effectively if it draws on the decades-long wealth of expertise, creativity, and problem-solving skills developed through the environmental justice movement and successful transitions seen in some former fossil fuel communities, rather than striving to tackle all interconnected problems simultaneously. When Pacala asked if there was a middle ground between technological solutions and massive social change, given that the seriousness of global warming warrants great urgency, Beckfield replied yes—pragmatist social mechanisms can be employed to solve real, day-to-day problems creatively and in ways that can scale up.

Jobs are also a critical social issue. Romero-Lankao stated that U.S. fossil fuel work is not being replaced with jobs of comparable quality—a serious problem that cannot be ignored in a just, equitable transition. Another participant agreed, noting that an *unjust* transition is already under way. To help workers feel an increased sense of agency and safety in their lives, the participant suggested that assistance programs should combat weak union power and labor laws by offering income, health insurance, education, and retraining opportunities for high-quality jobs. The participant cautioned that if decarbonization is pursued without justice, it can create new problems. In addition, Woodbury noted that a just transition must also help those communities burdened with a legacy of poverty, pollution, and powerlessness. Pacala agreed, adding that fossil fuel workers and disadvantaged communities are two very different populations, and a just transition will include separate policies to address their very different needs.

Learning Opportunities and Ideas

Krieger named the COVID-19 pandemic—in which critical failures resulted from an overemphasis on technical solutions like vaccines, paired with a lack of attention to and skill in addressing the social issues—as a cautionary tale of what happens when health justice concerns are not centered during a crisis. Instead, she said that technical and social problems must be resolved together—through meaningful engagement with other communities—to avoid pushing groups further apart. "We as humans, as people among other organisms on this planet, don't separately decide that this is a workplace issue ... this is a residential issue ... this is a transportation issue," Krieger said. "We embody all of that at once."

Cody TwoBears reminded participants that native communities are sovereign nations that do not have to wait for critical legislative votes; they can create their own laws and policies. One benefit of this is that they can recreate the success stories that surfaced during the workshop almost immediately and demonstrate that just transition is possible and worthwhile. Tribal nations are ready for this work, have land, and are connected to the U.S. power grid, but need support because they lack infrastructure to share information widely and spread these ideas. "We're ready, we're willing—I just wanted to shoot that out there to everybody to let you know that we just need the resources to be able to put these game plans in place," said TwoBears. "Let's show these examples in Native country so we can be a good example for the rest of the United States to follow."

Nock agreed that Indigenous regulatory structures are very different and pointed out that communities and structures across the United States also vary widely. In addition, different communities and people of color are not a monolith, and people experience a wide range of energy and affordability challenges. Science and engineering reports often push for objective studies, but she cautioned that seeking objectivity can also mask problems; for example, research demonstrates that the definitions of "affordable" and "vulnerable" are subjective.

A final idea raised by a participant was to push energy efficiency regulations onto manufacturers instead of low-income homeowners or renters, who should benefit from targeted programs and not be burdened by expensive regulations. This change could create momentum toward a more holistic shift in energy use, because energy is so central to people's lives. For example, utility access and affordability can have serious consequences for lower-income families, who could lose their children if they lose access to water or power. Pacala added that increasing the efficiency of lower-income homes was a significant but important task for the energy transition.

CLOSING REMARKS

Closing the workshop, Pacala reflected on several common themes that emerged from the breakout discussions. One key consideration is the need for more streamlined policies to address workforce issues, public health, and energy affordability challenges. For example, in some cases there may be opportunities to combine entry points for multiple programs into one application or criterion to reduce siloing, red tape, and fragmentation that communities frequently encounter when they try to solve multiple interconnected problems.

Another common theme was the need for more and better data—data to capture energy use, needs, and burdens from multiple angles and to understand how different technologies or policies might play out differently in various types of households, communities, and locations.

Finally, a critical theme underlying the whole workshop was the need for authentic stakeholder and community engagement, including in-person community visits to understand lived experiences and help community organizations build capacity, overcome barriers, and access resources. "There is no justice unless people have a say," said Pacala.

Appendixes

Appendix A

Statement of Task

Building off the needs identified at the Deployment of Deep Decarbonization Technologies workshop in July 2019, the National Academies of Sciences, Engineering, and Medicine will appoint an ad hoc consensus committee to assess the technological, policy, social, and behavioral dimensions to accelerate the decarbonization of the U.S. economy. The focus is on emission reduction and removal of CO_2, which is the largest driver of climate change and the greenhouse gas most intimately integrated into the U.S. economy and way of life. The scope of the study is necessarily broad and takes a systemic, cross-sector approach. The committee will summarize the status of technologies, policies, and societal factors needed for decarbonization and recommend research and policy needs. It will focus its findings and recommendations on near and mid-term (5–20 years) high-value policy improvements and research investments and approaches required to put the United States on a path to achieve long-term net zero emissions.

This consensus study will also provide the foundation for a larger National Academies' initiative on Deep Decarbonization. The committee will produce an interim report and a final report. The interim report will provide an assessment of no regrets policies, strategies, and research directions that provide benefits across a spectrum of low carbon futures. The final report will assess a wider spectrum of technological, policy, social, and behavioral dimensions of deep decarbonization and their interactions.

Specific questions that will be addressed in the final report include the following:

- Sectoral interactions and systems impacts—How do changes in one sector (e.g., transportation) impact other sectors (e.g., electric power) and what positive and negative systems level impacts arise through these interactions; how should the understanding of sectoral interactions impact choices related to technologies and policies?
- Technology research, development, and deployment at scale—What are the technological challenges and opportunities for achieving deep decarbonization, including in challenging activities like air travel and heavy processing; what research, development, and demonstration efforts can accelerate the technologies; how can financing and capital effectively support decarbonization; what are key metrics for tracking progress in deployment and scale up of technologies and key measurements for tracking emissions?
- Social, institutional, and behavioral dimensions—What are the societal, institutional, behavioral, and equity drivers and implications of deep decarbonization; how do the impacts of deep decarbonization differ across states, regions, and urban versus rural areas and how can equity issues be identified and the uneven distribution of impacts be addressed; and what is the role of the private sector in achieving emissions reductions, including companies influence on their external supply chains; what are the economic opportunities associated with deep decarbonization; and what are the workforce and human capital needs?
- Policy coordination and sequencing at local, state, and federal levels—What near-term policy developments at local, state, and federal levels are driving decarbonization; how can policies be sequenced to best achieve near, medium, and long-term goals; and what synergies exist between mitigation, adaptation, resilience, and economic development?

The study will coordinate with and leverage other efforts within the National Academies and outside groups that cover energy technologies, innovation, science, behavior, and policy. The committee will have expertise across engineering, policy, social and behavioral sciences, economics, and the physical sciences.

Appendix B

Workshop Agenda

JULY 26, 2022
WASHINGTON, DC

9:00 AM **Welcome and Workshop Overview**

9:15 **Session 1—Public Health, Safety, and Community Resilience**
Moderator: Michael Méndez, Committee Member

Panel:
Diana Hernández, Columbia University
Garvin Heath, National Renewable Energy Laboratory
Linda Rudolph, Public Health Institute
Jason Beckfield, Harvard University

Discussant: Nancy Krieger, Harvard School of Public Health

Session 2—Jobs and Workforce Development Opportunities
Moderator: Roxanne Johnson, Committee Member

Panel:
Mijin Cha, Occidental College
Madeline Janis, Jobs to Move America
Sarah Truitt, National Renewable Energy Laboratory

10:30 Break

10:45 **Session 3—Equitable Access to Transition Technologies, Infrastructure, and Programs**
Moderator: Paty Romero-Lankao, Committee Member

Panel:
Pjoy Chua, Los Angeles Department of Water and Power
Leslie Aguayo, Greenlining Institute
Darnell Grisby, Beneficial State Foundation
James Wilcox, New York State Energy and Research Development Authority

12:00 PM Lunch

1:00 **Session 4—Energy Affordability and Burdens**
Moderator: Adrienne Hollis, Committee Member

Panel:
Gregory Pierce, University of California, Los Angeles
Sacoby Wilson, University of Maryland
David Nahai, former Los Angeles Department of Water and Power
Destenie Nock, Carnegie Mellon University

2:15 Break

2:25 **Topical Roundtable Discussions**

Open for participation by in-person invited and public attendees, and remote invited attendees; closed to remote public viewing. Invited attendees and the committee will convene in small groups to discuss and develop specific needs and solutions to achieving the energy transition in a just and equitable way.

3:35 Break

3:45 **Pathways for the Energy Transition Discussion**

Open for participation by in-person and remote attendees; open to remote viewing. Representatives will present the key learnings from their small groups. The floor will then be opened to all audience members to ask questions, expand on the discussions, and connect ideas across different areas and experiences.

5:15 Adjourn Workshop